Lean Enterprise

The Complete Step-by-Step Startup Guide to Building a Lean Business Using Six Sigma, Kanban & 5s Methodologies

Table of Contents

Introduction

Congratulations on getting a copy of *Lean Enterprise: The Complete Step-by-Step Startup Guide to Building a Lean Business Using Six Sigma, Kanban & 5s Methodologies.* These days, it is more difficult than ever to build a business that can remain competitive in a world where customers can find your competition with just the click of a mouse. While there is only so much you can do when it comes to adjusting your profit margins, you can still find success by adjusting the method that will complete the processes in making your business successful.

Making a business Lean can give it the competitive advantage that the perpetual buyers' market takes away. However, it can be more difficult than it first appears which is why the following chapters will discuss everything you need to know in order to turn your business into a Lean mean fighting machine. First, you will learn all about the Lean system, its many benefits, and how you can get started creating your very own Lean system. Next, you will learn how to move the process forward in the right way by ensuring that you have the right goals in mind and that you go about implementing them in the best way possible.

From there, you will learn how to create a value stream

map which is vital when it comes to ensuring that your business's various processes are truly on point before learning how to choose the Lean system that best supports the flow of production. You will then learn about the importance of standardization before learning about the several important Lean tools which you can use to really whip your business into shape.

There are plenty of books on this subject on the market, thanks again for choosing this one! Every effort was made to ensure it is full of as much useful information as possible, please enjoy!

Chapter 1: Why Lean Matters to Your Enterprise

Lean principles were first discussed by an MIT student by the name of John Krafcik in his master's thesis. Before starting at MIT, Krafcik had already spent time as an engineer with both Toyota and GM, and he used what he learned from the Japanese manufacturing sector to posit a number of standards that he believed could help businesses of all shapes and sizes operate more efficiently.

The basic idea is that, regardless of what type of business a business is in, it is still just a group of interconnected processes. These interconnected processes can be categorized such as primary processes and secondary processes. The primary processes are those that directly create value for the business. Meanwhile, secondary processes are necessary to ensure the primary processes run smoothly. Regardless of the type of process you look at, you will find that they are all made up of a number of steps that can be carried out in a way that ensures they work as effectively as possible and that they need to be viewed as a whole in order for an effective analysis to be completed.

As a whole, you can think of the Lean process as a group

of useful tools that can be called upon to identify waste in the current paradigm either for the business as a whole or for its upcoming projects. Specific focus is also given to reducing costs and improving production whenever possible. This can be accomplished by identifying individual steps and then considering the ways they can be completed more effectively. Some tools that are commonly used in the process include:

- 5S value stream mapping
- error-proofing
- elimination of time batching
- restructuring of working cells
- control charts
- rank order clustering
- multi-process handling
- total productive maintenance
- mixed model processing
- single point scheduling
- single-minute exchange of die or SMED
- pull systems

Beyond these tools, Lean is also comprised of a number of principles that are loosely-connected around the twin ideas of the elimination of waste and the reduction of costs as much as possible. These include:

- flexibility
- automation
- visual control
- production flow
- continuous improvement
- load leveling
- waste minimization
- reliable quality and pull processing
- building relationships with suppliers

When used correctly, these principles will ultimately result in a dramatic increase in profitability. When given the opportunity, the Lean process strives to ensure the required items get to the required space in the required period of time. More importantly, however, it also works to ensure the ideal amount of items move as needed in order to achieve a stable workflow that can be altered as needed without creating excess waste.

This is most typically achieved via the tools listed above but still requires extreme buy-in at all levels of your organization if you ever hope for it to be effective in both the short and the long-term. Ultimately, the Lean system is only going to be as strong as the tools your company is using to implement it, and these tools will only ever be effective in situations where its values are expressed and understood.

Important principles

While it was originally developed with a focus on production and manufacturing, Lean proved to be so effective that it has since been adapted for use in virtually every type of business. Before adopting the Lean process, businesses have only two primary tenants. The first one focuses on the importance of incremental improvement while the other one is the respect for people both external and internal.

Incremental improvement: The idea behind the importance of continuous improvement is based on three principles. The first is known as the Genchi Genbutsu and is discussed in detail below. The second is known as Kaizen, and it has its own chapter later in this book. Finally, in order for continuous improvement to be truly effective, it is important to understand that you must lead your business with a clear knowledge of the challenges you are most likely to face as it is the only way to determine how to deal with them effectively.

When doing so, it is important to approach each challenge with the appropriate mindset which is one that supports the idea that every challenge leads to growth, which, in turn, leads to positive progress. Finally, you will also want to ensure that you take the time to challenge your preconceptions regularly as you will never know when your business might end up operating on an assumption that is no longer true. This is ultimately the best way to find unexpected waste which will ensure that you really start to improve internally not just in the short-term but in the long-term as well.

Respect for people: This tenant is both internal and external as it applies equally to your own people as well as to your customers. Respecting the customers means going the extra mile when it comes to considering their problems and listening to what they have to say. When it comes to respecting your team, a strong internal culture that is dedicated to the idea of teamwork is a must. This should further express itself in an implied commitment to improving the team as a means of improving the company as a whole.

Getting an edge

Prior to the digital age, businesses could determine their sales margin by starting with all the relevant costs, adding on a reasonable profit margin and calling it a day. Unfortunately, the prevalence of screens in today's society means that everyone is a bargain shopper, simply because it takes so little effort. This, in turn, means that you are not only competing against other businesses in your city, county or state, you are now competing with businesses all around the world as well. As such, there are only a few options when it comes to squeaking by with any profit margin whatsoever. Companies can either add additional real or perceived value, or they can reduce the amount of waste they are paying for as much as possible.

Most businesses find that it is better to determine their margins by looking at what customers are likely willing to pay for specific goods or service and then working backward from there. Ideally, you will be able to reduce that price by five percent to ensure you are truly competitive in a cost-conscious world. While it might not seem like much, this extra five percent is extremely important as customers are constantly on the lookout for the next sale, regardless of how much is actually being saved. The mental benefits that come along with being five percent better than those around them will be more than enough for them to commit to your product or service over all the rest.

Value add: Regardless of what your business does, you will find that there are Lean principles that can be implemented in order to improve the overall amount of value you are providing for your customers while also showing them you appreciate their business and respect them as individuals. What's more, you will be able to address the potential for waste in your organization while also maintaining flow and work to achieve perfection.

Often, you can manage this by doing something as simple as listening to your customers' specific wants and needs which will make it easier for you to determine what they really value the most when it comes to the niche your company habits. Value is most often generated by adding on something tangible that either improves or modifies the most common aspects of the good or service being provided. The goal is that this improvement is something the customer is willing to pay for, so when they receive it for free, they see this as a viable reason for your service to cost more out of the gate. It is also very important that the added value is very easy for the customer to claim because otherwise, they will feel that you have misled them.

Cost reduction: As the Lean system is already quite big on cutting down on waste in all of its forms, it should come as no surprise that it has some ideas when it comes to cost-reducing measures. For starters, it is important to understand that when it comes to Lean, all the different types of waste can be broken down into three types.

Muri is the name for the waste that forms when there is too much variation within common processes. Muda is the name given to seven different types of waste including:

- Transportation waste is formed when parts, materials or

information for a specific task are not available because the process for allocating resources for active products isn't where it should be.

- Waiting waste is created if a portion of the production chain has ideal time when they are not actively working on a task.

- Overproduction waste is common if the demand exceeds supply, and there is no plan in place to use this situation to the business's advantage. The Lean systems are designed to ensure that this number reaches zero so supply and demand are always in balance with one another.

- Defective waste is known to appear when some segment of the standard operating process generates some issue that needs to be sorted at some point down the line.

- Inventory waste is known to appear if the production chain ends up remaining idle between runs because it doesn't have the physical materials needed to be constantly running.

- Movement waste is generated when required parts, materials or information needs to be moved around successfully to complete a specific step in the process.

- Additional processing waste is generated if work is completed that does not generate value or adds value for the company.

- A commonly added eighth muda is the waste created by

the underutilization of your team. This can occur whenever any member of the team is placed in a position that doesn't utilize them to their full potential. It can also refer to the waste that occurs when team members have to perform tasks for which they have not been properly trained.

Muda also comes in three categories, the first of which is muda that doesn't directly add value but also cannot be easily removed if the system is going to continue working properly. When faced with this muda, the goal should be to work to minimize it slowly as a precursor to removing it completely. The second type of muda is that which has no real value, whatsoever, and you should work to remove it directly once you've become aware of it. Finally, the third type of muda doesn't directly add value but is required for regulatory purposes of one type or another. While it may be annoying, this type of muda is unavoidable in most instances and the best that you can do is ensure you are always updated to any relevant policies.

Chapter 2: Creating a Lean System

Lean leadership

With so much emphasis placed on improving efficiency, the Lean process naturally puts a lot of emphasis on team leaders who should be working hard to directly inspire their teams to adopt the Lean mindset. In the end, many Lean systems live and die by the leadership involved, which means it is important that those who are put in charge of leading the Lean transition are able to not only explain what's going on but are truly committed to the work that is being done as well. Some of the things that Lean leaders should strive to emphasize include:

Customer retention: When it comes to customer retention, Lean leaders need to take the time to consider not only what their customers want at the moment, but what they are likely to want in the future as well. Additionally, it is important to understand what a customer will accept, what they will enjoy, and what they will stop at nothing to obtain. The Lean leader should also work to truly understand the many ways the specific wants and needs of their target audience throughout the customer base.

Team improvement: In order to help their team members be their best, Lean leaders should always be available to help the team throughout the problem-solving process. At the same time, they are going to need to show restraint and refrain from going so far as to take control and just do things on their own. Their role in the process should be to focus on locating the required resources that allow the team to solve their own problems. Open-ended questions are a big part of this process as they will make it possible for the team to seek out a much wider variety of solutions.

Incremental improvement: One of the major duties of the Lean leader is to constantly evaluate different aspects of the team in order to ensure that it is operating at peak efficiency. The leader will also need to keep up to date on customer requirements, as this is something that is going to be constantly changing as well. Doing so is one of the only truly reliable ways of staying ahead of the curve by making it possible to streamline the overall direction of the company towards the processes that will achieve the best results.

In order to ensure that this is the case, the Lean leader will want to make time in their schedule to look at the results and then compare them to the costs as a means of discovering the best ways to use all the resources available to them at the given time. This will include things like evaluating the organization as a whole in hopes of making it more efficient and reliable. It will also involve evaluating the value stream to ensure that it satisfies the customer on both the macro and micro level.

Focus on sustained improvement: It is also the task of the Lean leader to ensure that improvements that are undertaken are seen through to the end as well. This will often include teaching the team members the correct Lean behaviors to use in a given situation and also approaching instances of failure as opportunities for improvement and innovation.

Three actuals

Lean leaders typically use a different leadership style than many of their peers, largely because being a Lean leader requires an understanding that the best way to analyze a situation is to physically be in the space where the situation occurred. Once there, the Lean leader needs to consider what is known as the three actuals, the broadest of which is known as Genchi and is the issue that led the leader to come to the place in question. Genbutsu is the idea that it is important to view what is being created or provided in action before making any moves. Finally, Genjitsu says it is always best to gather as much information as possible before making a decision one way or the other.

Creating a Lean system

In order to create a Lean system that lasts, the first thing you will need to do is consider the absolute simplest means of getting your product or service out to the public and put that system into effect. From there, you will need to continuously monitor the processes you have put in place to support your business in order to ensure that improvement breakthroughs happen from time to time. The last step is to then implement any improvements as you come across. While there are plenty of theories and tools that can help you do go on from there, the fact of the matter is that creating a Lean system really is that simple. Many of the chapters in this book will consist of deep dives on various tools that will make this process as easy to set up and as most likely to stick as possible.

There's more to business than profits: When using the Lean system, the end goal is to determine the many ways that it might be possible to improve the efficiency of your business. While an increase in profits is often a natural result of this process, this should not be the primary motivating factor behind undertaking a Lean transformation. Instead, it is important to focus on streamlining as much as possible, regardless of what the upfront cost is going to be since you can confidentially expect every dollar you spend to come back to you in savings.

There are limits to this, of course, and at a certain point, the gains won't be worth the costs. To determine where this line is, you can use a simple value curve to determine how the changes will likely affect your bottom line. A value curve is often used to compare various products or services based on many relevant factors as well as the data on hand at the moment. In this instance, creating one to show the difference between a pre- and post-Lean state should make such decisions far easier to make.

Treat tools as what they are: When many new companies switch to a Lean style of doing things, they find it easy to slip into the trap of taking tools to the extreme, to the point that they follow them with near-religious fervor. It is important to keep in mind that the Lean principles are ultimately just guidelines and any Lean tools you use are just that, tools which are there to help your company work more effectively. This means that if they need to be tweaked to better serve your team and your customers, then there is nothing stopping you from doing just that. Your team should understand from the very beginning the limits and purpose of the Lean tools they are being provided and, most importantly, understand that they are not laws.

Prepare to follow through: Even if you bring in a trained professional to help your team over the initial Lean learning curve, it will still ultimately fall to you, as the team leader, to ensure that the learned practices don't fall by the wayside as time goes on. It takes time to take new ways of doing things and turn them into habits, and it will be your job to keep everyone on until everything clicks and they start operating via the new system without thinking about it. Likewise, it is important that you make it clear just why the Lean process is good for the team as a whole and for the individual team member as if they are personally invested in it, then it is far more likely that they will stick with it, even if the going gets tough.

Chapter 3: Setting Lean Goals

In order to eventually make the right changes to your business, the first thing you need to do is ensure that you set the right goals. In order to make sure that your goals will put you on the right track, you need to ensure they are SMART which means they are specific, measurable, attainable, realistic and timely.

Specific: Good goals are specific which means you want to be sure that the goal you choose is extremely clear, especially when you are first starting out, as goals that are less well-defined are much easier to avoid doing in favor of activities that provide more positive stimulation in a shorter period of time. Keeping specific goals in mind will instead make it much easier for you to go ahead and power through whatever task you are currently undertaking.

When you aren't quite sure if the goal you have chosen is specific enough to actually improve your chance of changing for the better, you may be able to figure it out by running through the who, why, where, when, and how of the goal. Specifically, you are going to want to consider who is going to be involved with you when it comes to the completion of the goal? What exactly is it that is going to be accomplished? Where will it be taking place, why it is important that you ensure it is completed as quickly as possible, and how exactly you can expect to go about doing it. Once you can answer all five of the big questions, then you know you have a goal that is specific enough to generate the type of results that you are looking for.

Measurable: SMART goals are those which can be broken down into small, easily manageable chunks that can be tackled one piece at a time. A measurable goal should make it easy to determine when exactly you are headed off track so that you can self-correct as quickly as possible. Measuring your progress will make it easier for you to keep up the good work.

Attainable: Perhaps more important than anything else, if a goal that you set is unattainable, especially the first goal that you set using this system, then you are going to unknowingly be wasting valuable time and energy while creating negative patterns that end in failure. What's more, you will end up reinforcing fixed mindset ideals, making this a bad choice any way that you look at it. This means that when it comes to setting goals, you are going to want to have a clear understanding of the current situation and anything going on with the business that would make it less likely to succeed as far as that goal is concerned.

Realistic: A good goal is one that is realistic, in addition to being attainable, which means that you can expect success without something extremely unlikely being required to push reality into your favor. An ideal goal is one that is going to require a good amount of work to achieve, while still remaining not too difficult as to become unrealistic. Additionally, you are going to want to shy away from goals that you can meet without putting any real amount of effort as goals that are too easy can actually be demotivating as it then becomes easy to continue putting them off until they eventually fade into oblivion.

Timely: Studies show that the human mind is more likely to actively engage in problem-solving behaviors when there is a time limit involved in the successful completion of the task in question. What this means for the goals you are setting is that if you have a firm completion date in mind for when you want to have reached your goal, then you will work harder in the period leading up to that date. This means that you are going to want to pick completion dates that are strict enough to truly motivate you to do whatever it is you have in mind, while at the same time, not being so strict that there is no realistic way that you can complete the task on time. The goal here is to throw a little extra hustle into your step and not force you to keep a grueling schedule, thus, ensure that you can always meet the schedule you set for the best results.

Policy Deployment

Also known by the name Hoshin Kanri, policy deployment is a way of ensuring any SMART goals that are set at the management level ultimately filter down to the rest of the team in a measurable way. Making proper use of policy deployment will essentially ensure that anything you are planning to put into effect doesn't accidentally end up creating more problems than it ultimately solves. It will also help to ensure as little waste as possible is generated as a result of things like inconsistent messaging from management or all around poor communication. The goal in this instance should not be to force various team members into acting in a specific way. It is about generating the type of vision for the business that everyone can appreciate and understand how it pertains to both the team and the customers.

Implementing the plan: Once all of the relevant SMART goals have been finalized, the next thing you will want to do is to group them together based on which members of the team will ultimately be tasked with solving them. Keep in mind that the fewer number of goals, the more likely it is that they will be acted upon in a reasonable timeframe. If your goals cannot be generalized in such a way, it is important to instead begin with the ones that are sure to make the biggest difference overall and work down the list from there.

Regardless of what goals you ultimately settle on, it is important that you take special care to ensure that there are no goals that do not have one person specifically assigned instructions to keep tabs on the overall progress while providing status reports when needed. This person should also be someone who can be counted on to make it clear to the other team members how important the goal is for the business as a whole and how it will make things easier in the long run.

Consider your tactics: Those who will be responsible for making the goal a reality should, in turn, be the ones who decide how the goal can best be completed by the team as a whole. However, this process should still include back and forth interaction between all levels of the team just to ensure that the tactics and the goal align properly. Tactics are likely to change as the goal heads towards success, which means it should be studied from time to time to ensure they remain appropriate for the goal in question.

Moving forward: Once the tactics have been agreed upon by all parties, it will then be time to actually put them into practice. This will be the stage where the team can really take over, though quality goals should still require buy-in from relevant parties. During this period it is important to ensure all communication from management is on message, to properly ensure that actions and broader goals will continue to align.

Review from time to time: It is important to keep in mind that once the action is in progress, the team leader will need to change the action as needed. This means that they will also be monitoring things as they proceed, hopefully, according to plan. Remember, Lean systems are always being improved upon, which means your goals and their implementation should be no different.

Chapter 4: Simplifying Lean as Much as Possible

All of the products and services that are generated by your business have a mixture of three different value streams that can ultimately be used productively if you take the time to understand them fully. These include the concept to launch stream, the creation of customer stream and the order to customer stream. In order to ensure you are getting the greatest overall value out of all of the processes your business finishes, it is important to look at a value stream map as it is an excellent way to ensure you are maximizing efficiency at every turn.

The average value stream map will include everything that ultimately comes together to generate value for the customer including activities, people, materials, and information. To properly visualize a value stream, you will want to follow the Plan/Do/Study/Act process, also known as the Lean cycle. To get started, you will want to plan out the task ahead by focusing on one goal at a time. From there, you will want to make a list of everything that will need to be done in order to ensure the task is finished successfully. This is then followed by the step of following through, the results being studied, and acted upon as required.

Create your own value stream map

A properly constructed value stream map is a vital part of the process as it will allow you to see the big picture by mapping out the entire flow of resources from their disparate starting points all the way through when they come together and eventually make it into the hands of the customer. As such, it then makes it far more of a manageable task to determine the points in the process that are bottlenecking the overall efficiency of your business's process and thus, taking the first steps towards adopting Lean processes.

While one person can certainly work through the following steps, the value stream maps that prove to be the most effective are often those created by the entire team, so that those who are the most knowledgeable about each step will be able to give their two cents as well. Your initial value stream map should be thought of as a very rough draft and should be constructed as such, which means planning it out in pencil and expecting lots of rewriting as you go along.

Consider the process: The first step in this process will be to consider exactly what it is you will be mapping. For businesses that are first starting out with the Lean system, you will want to begin by considering the various processes that are ultimately going to prove to be of the greatest value to the team as a whole and then work down the list from there. If you still can't decide where to start, then you will want to turn to your customers, consider what they have to say and start with the areas where you regularly receive the most complaints.

What is known as a pareto analysis is an effective tool at this juncture as it can make it easier for you to find the right place to start if you aren't sure where your efforts will be best put to use. It is a statistical analysis technique that can prove especially useful if you are looking at a few different tasks that are sure to generate serious results if only you could decide which one to use first. The goal, in this case, is to focus on the 20 percent of your business that, if nurtured, could ultimately generate 80 percent of your total results. Your initial value stream map may focus on only a single service or product or on multiples that share a significant portion of the process.

Choose your shorthand: The symbols you use to denote various stages of the process you are mapping don't really have any hard and firm guidelines as they are going to be unique to every project and every business. Regardless of what you and your team ultimately choose, it is important to create a list of all of the symbols you are using and what they mean so that anyone who comes in after the fact can easily get caught up. From there, it is important to stick to the designated symbols and not make anything up on the fly. Additionally, if the business is working on more than one value stream map at a time, it is important that the symbols correspond between the two. Otherwise, things can quickly spiral into illegibility.

Set limits: If taken from a broad enough scope, virtually every value stream map for your business can be connected to other value stream maps or go into greater detail. At some point, however, this is going to be counterproductive and you will have to set limits on what the value stream map is going to account for if you ever hope to successfully move forward. Likewise, if you let this part of the process get out of hand, then the map can lose focus and become less useful as a result.

Start with steps that are clearly defined: After you have a clear beginning and end for the process you are mapping, the next thing to do is to make a list of all of the logical steps that need to be taken from start to finish. This shouldn't be an in-depth look at every link in the chain, but instead, should be an overview of the major stages that will need to be looked at more closely as the process moves towards completion.

Consider the flow of information: One important step in the value stream mapping process that sets it apart from other similar mapping processes is that each value stream map also accounts for the way that information flows throughout the process from beginning to end. What's more, it will also chart the way information passes between team members as well. You will also need to ensure it takes into account the ways the customer interacts with your business, in addition to how frequently such interactions occur. You will also need to ensure the communication chain includes any suppliers or any other third parties the company deals with.

Further details: When it comes to breaking the process down to its most granular level, you may want to include a flow chart with your value stream map as well. A flow chart is a great way to map out the innermost details of how a given process reaches completion. This is also an excellent way to determine the types of muda you are dealing with, so you can consider if they can be removed from the process.

If you are interested in considering the ways your team physically moves around your space, then a string diagram can also prove effective. To generate this type of diagram, you will map out your business's workspace by drawing in what each member of the team has to do and where they have to go in order to fully complete the process. You will want to draw different team members or different teams in different colors to keep things from getting too confusing. From there, charting the flow of information as it relates to this data can lead to surprising conclusions regarding flaws that might otherwise go unnoticed for years.

Collecting data: When it comes to outlining your initial map of a value stream, you may find yourself with certain aspects of the process that require additional data before anything can be determined with any real degree of certainty. The data that you may need to track down will include:

- cycle time
- total inventory on hand
- availability of the service
- transition time
- uptime
- number of shifts required to complete the process
- total available working time

When it comes to collecting this data, it is important to always remember to go to the source directly and find the details you are looking for rather than making assumptions. Furthermore, it is important to get the most updated numbers possible as opposed to looking at older, more readily available figures or hypothetical benchmarks. This may mean something as hands-on as physically keeping an eye on every part of the process in question so you can take relevant notes.

Watch the inventory: Even if you are relatively certain about any inventory requirements for the process in question, it is vital that you double check before you commit anything to the value stream map. Minor miscalculations at this point could dramatically skew your overall results and essentially nullify all of your hard work if you aren't careful. This means you definitely need to adopt a measure twice in order to see the best results. After all, inventory is prone to building up for a wide variety of reasons and there is a good chance that you won't know it until you take a closer look and do a once over on what's really on hand. You can also use this step as an excuse to take stock of exactly what the team is working with and determine how far it will actually stretch effectively.

Using the data: After you have finished visualizing the steps found in your most important process, you are now ready to use it as a means of determining where any problem points might be. You will especially want to keep an eye out for processes that include redoing any previously completed work, anything that requires an extended period of resetting before work can begin again, or long gaps where parts of the team can do nothing except wait for someone else to finish, those that take up more resources than your research indicates you should or even just those that seem to take longer than they should for no particular reason.

Generate the ideal version of the value stream: After you have determined where the bottlenecks are occurring, you will want to create an updated value stream map that represents how you want the process to proceed once you have everything properly sorted. This will provide you with an A to C scenario, where figuring out the pain points represents B. Ideally, it will also provide a clue as to how you can go about eliminating the waste from the process in order to create an idea, which you can really strive for both in the short and the long-term.

Once you have determined the ideal state for the process, you can work out a future value stream map that will serve as a plan on how to take the team from where you are currently to where you need to be. This type of plan is often broken down into sections that last a few months, depending on what needs to be done. Additionally, most future value stream maps will come with multiple iterations because they will need to change several times as the project nears completion.

When working through various variations of the value stream map, it is important to pay close attention to the lead time available for various processes. The lead time is the amount of time it will take to complete a given task in the process and, if not utilized as efficiently as possible, it can easily lead to a wealth of bottlenecks. Remember, when it comes to creating the best value stream map possible, no part of the process is beyond scrutiny.

Chapter 5: Lean and Production

When putting together a Lean system for your business, you will eventually determine where the waste is hiding in your current processes, which is when it will be time to consider what can be done about the flow of the process. Often, the answer will come in either the form of a continuous flow model or a batch model.

Continuous flow model: The ideal version of the continuous flow model sees the customer order a product or service before the necessary steps are taken to generate the product or service that is being paid for. The product or service is then delivered to the customer who then pays for their order. The end result here is that there is no downtime between when the customer puts in their request and when it is completed. Furthermore, every step is going to smoothly flow into the next as a means of ensuring that overall downtime is reduced as much as possible.

The biggest upside to the continuous flow model is that it allows business to make assumptions and plan for the future based on a profit level that prioritizes continuity and stability. A continuous flow setup also features less waste than other types of processes. The biggest downside, however, is that this type of scenario can be difficult to produce reliably as every step in the process is rarely equal, regardless of how clear the value stream map might be. If you are striving to create a continuous process scenario, then you should be aware that new problems can also appear quickly if your available margin for error begins to shrink.

In order to persist despite these drawbacks, you will want to do your best to attack these problems head-on and be determined to push through them if you hope to find success. Additionally, if you hope to choose this route, it is important to start your journey to a Lean system with this in mind, as a continuous system is only going to work if every part of the system is completely in sync with all the rest.

Heijunka is a useful tool when it comes to facilitating this process as it promotes leveling out the quantity and quality of the process over a prolonged period of time in hopes of making everything as efficient as possible and, what's more, to expressly prevent batching. While it might sound complicated, in reality, this process can be as simple as making sure your team has all of the storage space they need to organize the various parts of the project. They store them in folders that are organized based both on the frequency of use as well as the due date. Folders that are currently in use can be stacked vertically on top of one another while those that are idle can be stored horizontally out of the way somewhere. There are also numerous other types of organizational methods that promote heijunka, so it can be helpful to explore them all to see what offers the most benefit to your business.

Batch production: Unlike with the continuous flow model, with the batch production model, the steps in the process to create the product or service are all completed in bulk one after the next. This makes it the superior choice when it comes to situations where what the process generates is evergreen as this will allow it to be stockpiled as a direct counter to erratic customer demand. Depending on the specifics of your business, batch production can also dramatically decrease the associated production costs as few team members can move from step to step instead of having all the steps operating at once. It also provides lots of opportunities when it comes to cross-training.

As a general rule, you can count on batch production to be less productive when there are a greater number of individual steps that are required to complete the process. This is because the starting and stopping times would need to be calculated for each which can add up quickly if batch sizes are quite large. This can also potentially create a delay if a customer places a large customized order when a batch is already in the middle of production.

Takt time

Short for the word Taktzeit, Takt time is a variation of the Japanese principle of measuring time, despite its German name. Despite the fact that it is primarily used in production environments, it can have a beneficial effect on most tasks performed in a business environment as well. Specifically, Takt time is the time it takes for a team to start a new process after completely finishing the last, assuming the production rate is equal to the rate of customer demand.

Determining takt time: If your team completes processes one at a time throughout the workday, the takt time of that process can then be determined by taking the time that has elapsed between two processes, assuming of course that demand is still being met. This means it can be written as $T = T_a/D$. In this case, T is your Takt time, T_a is the amount of time available to finish processes, and D is the amount of demand that the process experiences.

You will not want to automatically take these results as fact, however, as it is rare to find a team that can run at peak efficiency at all times. As such, when it comes to determining an accurate takt time, you will want to add in some wiggle room here to compensate for the fact. From there, you will want to adjust your takt time based on additional customer requirements or team demands.

Takt time benefits: Once you have determined the accurate takt time for your business's processes, you will find that a number of additional beneficial options open up to you. First and foremost, you will find that it is clear which steps in the process are the bottlenecks which will make it easier to take steps to mitigate them specifically. Likewise, if you have any processes that typically go off the rails, that problem will be made apparent as well.

As a general rule, takt time places additional emphasis on steps that add value to the process as a whole, which makes it easier to use if you are looking for muda to remove from the process. What's more, once the team gets used to the concept of takt time, they will find that it is much easier to track how productive they are being throughout the day.

Be aware: Takt time is not a set-it and forget-it type of affair which means that if you find that your demand changes dramatically, then you will need to recalibrate all of your takt time to adjust for this fact. This also means that if your demand isn't relatively stable, then determining your takt time might not be very beneficial on its own. If you try and force your process into a takt timetable and it isn't a good fit, then all you will end up doing is causing even more waste in the long run.

Likewise, you will need to be aware of the way in which

the products or services produced by your processes fit together or else, you risk creating bottlenecks anyway which will throw off the accuracy of your takt time. As a general rule, the shorter the takt time, the greater the amount of strain that resources including both machinery and people will experience on a regular basis.

Chapter 6: Run a Lean Office

One of the truly great parts of the Lean system is the potential it holds when it comes to standardization, specifically when it comes to minimizing waste. Much like when it comes to setting goals, setting work parameters that are clearly standardized makes it easier to answer specific questions about the process. This should include things like who will follow through on the process once it has been outlined, how many people will it take, what will the end result be, what the metrics for success should be, what is required to meet them, how long the process will take and more. These are all questions that ultimately need to be asked in order to guarantee your standardization measures don't end up creating new problems instead of solving existing ones.

Workflow standardization is not expressly designed to ensure that processes are completed as quickly as possible. Rather, it is about utilizing the most effective practices possible in order to ensure they are completed with the same level of reliable quality each and every time. You will also do well to remember that standard practices will naturally change over time as improvements to safety, quality, and productivity are found. You will want to take care to avoid becoming so reliant on a single type of standardization that you end up actually allowing it to hold you back from future progress.

With that being said, it is also important to avoid falling into the trap of undertaking a round of standardization solely for standardization's sake. Instead, it is important to consider if standardization is really the right choice by considering the various processes already in place and asking yourself if they would be of a higher quality, performed to a higher safety standard or completed with less waste. If you move forward, and this is not the case, then all you are doing is inviting in waste.

Furthermore, standardization should involve more than simple instructional documents. It should be created from the input of those who perform the processes on the regular and then combined with a fresh round of customer feedback to ensure all bases are covered. The reasons for the standardization process should be clear to everyone involved before getting started for the best results.

KPIs

KPIs, also known as key performance indicators, are extremely useful when it comes to determining the ideal steps to take during the standardization process. KPIs are also useful when it comes to measuring the overall success of the process as a whole based on numerous different metrics. Choosing the right KPI to focus on is a matter of considering what metrics you value most at the moment as well as in the long-term. There are a wide variety of indicators to choose from, all of which are useful in different circumstances and when it comes to accurately defining specific values. Essentially, each KPI can be considered an object which is useful in various value-add scenarios.

Choosing indicators: When it comes to identifying the KPIs you want to use for your business, the first thing you will need to do is ensure that your process is already well-defined as this will help you have a true handle on the specifics of every aspect of the process as well as the best ways to determine the ideal means of completion. It is important to only stick with indicators that are relevant to the goal you are currently working towards. Otherwise, they can easily be altered dramatically by factors that are literally outside of your control.

Much like with your goals, it is important that your KPIs are SMART and that they clearly indicate specific information for a specific purpose. You will also want to choose options that are easily measured while still providing accurate results if at all possible. Much like goals, KPIs are useless if they are not realistically achievable. The most effective KPIs are those that are relevant to the success of the business in the moment or in the future while also including an element of time that has specific periods as they relate to the data.

Be aware: It is also important to keep in mind that while determining specific KPIs isn't too difficult, keeping track and compiling the relevant data can be more difficult than it first appears. Furthermore, additional values, including those for things such as staff morale, are difficult to gauge accurately. Before you invest resources into generating KPIs, it is important to first make sure they are adequately measurable and useful. Otherwise, you will be on your way to creating even more waste.

You will also need to ensure the focus remains on keeping the KPIs on the data that they are detailing and use it as a means of determining the overall health of the business as opposed to a set of numbers that can only move one way. If your team ends up too focused on reaching a predetermined KPI, the data they return will be biased and inaccurate.

Chapter 7: Kanban

Kanban is a method of scheduling that is often used once a Lean system has been put in place. It serves as a type of inventory management whose end goal is to minimize waste in the supply chain. It also tends to come in handy when it comes to pinpointing problems as it makes these problem areas stand out more than they otherwise would. You will also find it useful when it comes to locating the upper end of work related to inventory that is currently underway to ensure the process doesn't overload.

This is a demand-driven system which means, it is often implemented as a means of ensuring quicker turnaround times while at the same time limiting the required inventory and increasing the overall level of competitiveness between the implementation team. It was first put into effect by Toyota in the 1940s after the company performed a study on supermarkets and decided to use similar practices in order to keep their factories optimally stocked at all times. This eventually became one of the core Kanban ideas of keeping inventory amounts level with consumption rates. Additional supplies are then added based on a predetermined set of signals to ensure that stock remains near the ideal level at all times instead of dipping low or overflowing at certain points. The signals in question are all based upon customer signals which means they can change at the moment if needed.

Kanban rules

- Each process creates an amount set by the Kanban.
- Following processes collects the number of items set by the Kanban.
- Nothing is created or moved with a Kanban.
- Kanban is attached to related goods.
- Defective products are not counted in the Kanban.
- The fewer the Kanban, the more sensitive the system is.

Kanban cards: Kanban cards are the means by which signals are used to keep the entire team on track as it moves through the process. While they were actual cards when the system was created, these days, there is a wide variety of software out there that will provide the relevant signals without bringing physical cards into the process. Kanban cards generally represent consumption via a lack of cards in one area which, by necessity, drives another part of the process to do what needs to be done in order to pass the relevant cards along.

These days, the electronic Kanban system is even more effective than its physical predecessor, making it a sure thing to ensure that cards get where they need to be when they need to be there. These systems often mark set types of inventory with specific barcodes that are then scanned throughout the process. Each scan then sends a specific message to the Kanban program which routes it as needed.

Kanban types: There are two main varieties of Kanban systems namely production systems and transportation systems. Production systems are sent as a means of authorizing production or a specific number of items, while the transportation systems are used as a means of authorizing the movement of specific items once they have been created.

Three bin system: An example of a basic type of kanban system is the simple three bin system for the supplied parts in scenarios where manufacturing does not take place in-house. One bin represents the factory floor (or the primary point of demand anywhere else), the second bin represents the factory store (the control point for the inventory), and the final bin represents the supplier. The bins then can have removable cards containing relevant product details along with any other important information.

When the factory floor bin empties out because the relevant parts were all taken up by various parts of the manufacturing process, the empty bin, and thus its kanban card, are then returned to the factory store (also known as the inventory control point). The factory store then replaces the empty factory floor with the full factory store bin which also contains its own kanban card. The factory store then sends the empty bin and its related kanban card on to a supplier. This, in turn, causes the full product bin from the supplier to eventually replace the empty bin on the factory floor and the process starts all over again. Thus, the process never runs out of product. This could also be described as a closed loop, since it provides the exact amount required, with only one spare bin so there is never oversupply. This 'spare' bin allows for uncertainties in supply, use, and transport in the inventory system. A good kanban system calculates just enough kanban cards for each product. Most factories that use kanban use the colored board system

Chapter 8: 5s

When it comes to determining what wasteful processes you are dealing with, it is important to ensure the work environment is in optimum shape for the best results. The 5S organizational methodology is one commonly used system based around a number of Japanese words that, when taken together, are first rate when it comes to improving efficiency and effectiveness by clearly identifying and storing items in their designated space each and every time.

The goal here is to allow for standardization across a variety of processes which will ultimately generate significant time savings in the long-term. The reason it is so effective is that each time the human eye tracks across a messy workspace, it takes a fraction of a second to locate what it is looking for and process everything around it. While this might not be much if it happens now and then, if it is happening constantly across an entire team, then it can add up to serious time loss when taken across the sum total of the process in question

Sorting: Sorting is all about doing what can be done in order to always keep the workplace clean of anything that isn't required. When sorting, it is important to organize the space in such a way that it removes anything that would create an obstacle towards the completion of the task at hand. You will want to ensure that process-critical items all have a unique space that is labeled as well as a space that is designated for those things that simply don't fit anywhere else. Moving forward, this will make it easier to keep the space free of new distractions. Nevertheless, it will still be important to encourage team members to prune their personal space regularly to keep new obstacles from popping up.

Set in order: When it comes to organizing the items in the workspace themselves, it is important to ensure all the items are organized in the order that they will most likely be used. While doing so, it is important to take care to ensure that everything required for the most common steps remains readily at hand to reduce movement waste as much as possible. Over time, keeping things in the same place will ensure that the process can be completed faster each time as muscle memory takes over, and team members are able to reach for things without looking for them.

It is important to keep an open mind during this step since ensuring that the workspace is set up in such a way that ease of workflow is promoted may require more than a simple organization, it may require a serious rework of existing facilities. Additionally, ensuring everything is arranged correctly will make it easier for you to create steps for each part of the process that anyone new to that part of the process can follow.

Shine: Keeping the workspace clean is an essential part of maintaining the most effective workspace possible. It is important to emphasize the importance of daily cleaning both for the overall efficiency boost and its ability to ensure that everything is where it is supposed to be so that there are no issues the next time they are needed. This will also provide an opportunity to have a regular maintenance if any is needed, which will serve to make the office a safer place for everyone. The end goal should be that any member of the team should be able to enter a new space and understand where the key items are located in less than five minutes.

Standardize: The standardize step is all about making sure the organizational process itself is organized in such a way that it can be applied throughout the entire business structure. This will make it easier to maintain order when things get hectic and also ensure that everyone can be held to the same reliable standard.

Sustain: Sustaining the process is vital as taking a week or more to properly get everything in order only to have it all fall apart six months later is going to accomplish nothing in the long-term. As such, it is important to ensure that the organization is a vital part of the DNA of the business in moving forward. If things are truly sustainable in this regard, then team members will be able to successfully move through the process without expressly being asked to. Unfortunately, you won't be able to expect this type of sustainability overnight. It will require plenty of training and an adoption of the idea as part of the business's culture.

Great starter tool: If your plan for your business is to transition to additional advanced Lean concepts over time, then 5S is a great way to start moving employees in that direction. It is especially effective with employees who are extremely stuck in their ways as, once they initially get on board, they will be hard pressed to deny the benefits in completion times that come with the improved organizational version. This, in turn, will make it easier for them to get on board with additional changes that may come in the future.

As a rule, when rolling out a new system like this, you can expect team members to only care about two things, the way the new system is going to affect them specifically and if the Lean process has actually seen results. This is also what makes 5S a great starting point as it has easily understandable answers for each that anyone can understand once they see the first workspace transformed for efficiency.

Knowing if 5s is right for your business: While 5S is a great choice for some businesses, it is not a one-size-fits-all solution, which means it is important to understand both of its strengths and its weaknesses when moving forward. Perhaps its biggest strength is that when implemented successfully, it is sure to help your team define their processes more easily while also helping them claim more ownership of the processes they are associated with as well. This extra structure also has the potential to lead to a much greater degree of personal responsibility among team members which will lead to a greater feeling of accountability throughout the process. When everything goes according to plan, this will then lead to further improved performance and better working conditions for everyone involved.

What's more, implementing 5S also has the potential to more likely make long-term employee contributions thanks to an internalized sense of improvement. Ideally, this will continue until the idea of continuous improvement becomes the order of the day. When done correctly, using 5S will also provide further insight into the realm of value analysis, equipment reliability, and work standardization.

On the other hand, the biggest weakness of 5S is that if it, and its purpose, are not communicated properly, then team members can make the mistake of seeing it as the end goal and not a means to an end. 5S should be the flagbearer for success to come in the future, not the sum total of a company's journey into Lean processes. Specifically, businesses whose movement is constrained significantly by external factors will have a hard time using 5S, and companies that currently have a storage problem would do well to solve it before attempting a 5S transition.

Additionally, just because 5S is a great fit for many companies doesn't mean that it will be the best choice for your team. This is especially true for smaller teams or for teams where team members wear many hats. Just because it is a popular way to implement Lean principles doesn't mean that it is going to be right for everyone. Moving ahead anyway and enforcing organization simply for the sake of organization won't do much of anything when it comes to generating real results. Instead, it will only generate new waste and it will only continue to do so before it is abandoned entirely.

This is especially true for businesses that run on a wide variety of human interaction, various management styles, and other management tools. However, when the various aspects work together properly, they will actually end up generating extra value for the customer which is a vital part of any successful business. If you blindly press forward with a 5S mentality, however, then it can become easy to lose sight of the outcome for the customer in pursuit of a perfect outcome or a perfect implementation of 5S principles.

Above all else, when implementing 5S, it is important that you stress to your team that 5S is something that should be part of the natural work routine and standard best practices, not an additional task to be done outside of daily work. The goal of 5S is to enhance the effectiveness of the workflow at every step in the process. Separating out the 5S into its own separate layer is the complete opposite of what the process stands for.

Chapter 9: Six Sigma

Six Sigma is the shorthand name given to a system of measuring quality with a goal of getting as close to perfection as possible. A company operating in perfect synchronicity with Six Sigma would generate as few as 3.4 defects per million attempts at a given process. Zshift is the name given to the available deviations between a process that has been completed poorly and one that has been completed perfectly.

The standard Z-shift is one with a number of 4.5, while the ultimate value is a 6. Processes that have not been viewed through the Six Sigma lens typically earn around a 1.5.

Zshift Levels: A Six Sigma level of 1 means that your customers will get what they expect roughly 30 percent of the time. A Six Sigma level of 2 means that roughly 70 percent of the time, your customers will get what they expect. A Six Sigma level of 3 means that roughly 93 percent of the time, your customers will be satisfied. A Six Sigma level of 4 means that your customers will be satisfied more than 99 percent of the time. A Six Sigma level of 5 or 6 indicates a satisfaction percentage of even closer to 100 percent.

Six Sigma Certification Levels: Six Sigma is broken into numerous certification levels depending on the amount of knowledge the person in question has regarding the Six Sigma system. The executive level is made up of management team members who are in charge of actively setting up Six Sigma in your company. A Champion in Six Sigma is someone who can lead projects and be the voice of those projects specifically.

White belts are the rank-and-file workers; they have an understanding of Six Sigma, but it is limited. Yellow belts are active members on Six Sigma project teams who are allowed to determine improvements in some areas. Green belts are those who work with black belts on high-level projects while also running their own yellow belt projects. Black belts lead high-level projects while mentoring and supporting those at other tiers. Master black belts are those who are typically brought in specifically to implement Six Sigma and can mentor and teach anyone at any level.

Implementation: Giving your team a compelling reason to try Six Sigma is vital to the overall success of the process. In order to ensure that Six Sigma is properly implemented, it is important that you properly motivate your team by explaining how crucial the adoption of a new methodology really is. The most common choice in these situations is to create what is known as a burning platform scenario.

A burning platform is a motivational tactic wherein you explain that the situation the company now finds itself in is so dire (like standing on a burning platform) that only by implementing Six Sigma is there any chance of long-term survival for the company. Having stats that back up your assertions is helpful, though, if times aren't really so tough, a bit of exaggeration never hurt. Adapting to Six Sigma can be difficult, especially for older employees and a little external motivation can make the change more palatable.

Ensure the tools for self-improvement are readily available: Once the initial round of training regarding Six Sigma has been completed, it is important that you have a strong mentorship program in play while also making additional refresher materials readily available to those who need them. The worst thing that can happen at this point is for a team member who is confused about one of the finer points of Six Sigma to try and find additional answers only to be rebuffed due to lack of resources.

Not only will they walk away still confused, but they will also be rebuffed for trying and not rewarded for taking an interest in the subject matter. A team member who cannot easily find answers to their questions is a team member who will not follow Six Sigma processes when it really counts.

Key principles: Lean Six Sigma works based on the common acceptance of five laws. The first is the law of the market which means that the customer needs to be considered first before any decision is made. The second is the law of flexibility wherein the best processes are those that can be used for the greatest number of disparate functions. The third is the law of focus which states that a business should only focus on the problem the business is having as opposed to the business itself. The fourth is the law of velocity which says that the greater the number of steps in a process, the less efficient it is. Finally, the last is the law of complexity which says that simpler processes are always superior to more complicated ones.

Choosing the best process: When it comes to deciding what process to apply the Six Sigma treatment to, the best place to start is with any processes that are already defective and need work to reduce the number of times they occur. From there, it will simply be a matter of looking for instances where takt time is out of whack before looking into those steps where the number of available resources can be reduced as well.

Methodologies: There are two main ways to get the most out of Six Sigma, DMADV and DMAIC.

DMAIC is an acronym that is useful when it comes to remembering five phases that can be useful when it comes to creating new processes.

- Define what the process should do based on customer input.
- Measure the parameters that the process will adhere to and ensure it is being created properly by gathering relative information.
- Analyze the information you have gathered.
- Improve the process based on the analysis you have completed.
- Control the process as much as you can by finding ways to reliably decrease the appearance of delinquent variations.

DMADV, on the other hand, also has five phases that correspond to the DMAIC phases.

- Define the solutions the process should be providing.
- Measure the specifics of the process to determine its parameters.
- Analyze the data you have collected up to this point.
- Design the new process using your analysis.
- Verify the results as needed.

Deciding if Six Sigma is the right choice: While the Six Sigma system has something to offer teams of all shapes and sizes, that doesn't mean that it is going to be the best fit for all of them. This is especially true as implementing it successfully depends on numerous different specifics, starting with the conviction of those who are looking to implement the system in the first place as well as the company's overall culture. This is why it is best to start with something less high-impact like 5S as a way to ease your team into things that are more of an overall change like Six Sigma.

When deciding if a Six Sigma transition is feasible, it is important to ensure that it is not seen as a fad and can instead be seen as an evolution of the ideals already in place. Generally speaking, the more involved the team leadership is from the beginning, the more onboard the rest of the team will be as well. It is important that the company culture is perceived to be one that supports this sort of positive change and to remember that if the management team can't come to a consensus on the new program, then it is sure to be dead in the water. This doesn't mean that absolutely every member of the team needs to be committed to the idea of Six Sigma from the start, but it does mean that the change needs to be institutional so the public front always needs to appear united.

After all, Six Sigma was founded on the idea of leaders mentoring those beneath them in order to ensure Six Sigma works as it should, but this need to be a full-time job for some people, at least until the new habits start to solidify among the team as a whole. Once this occurs, you can then count on the team members to keep one another on track. To ensure they get to this point, you are going to want to let them know that their progress is being tracked so that every team member constantly feels as though they are improving up until the point where they internalize the Six Sigma principles.

Conclusion

Thank you for making it through to the end of *Lean Enterprise: The Complete Step-by-Step Startup Guide to Building a Lean Business Using Six Sigma, Kanban & 5s Methodologies*. Let's hope it was informative and able to provide you with all of the tools you need to achieve your Lean implementation goals. Just because you've finished this book doesn't mean there is nothing left to learn on the topic. Expanding your horizons is the only way to find the mastery you seek.

When it comes to implementing Lean techniques successfully, it is important to be realistic when it comes to the timeframe required to not just ensure the entire team is up to speed, but that they have internalized the core Lean principles you are trying to instill. You will need to take a long hard look at your team and your business as a whole and decide where the most work is going to need to take place. Every business has limited resources, after all. It is important to think wisely prior to allocating them.

While you can easily get sucked into a pattern of changing everything, in order to ensure your business really is as Lean as possible, you should keep in mind that discretion is the better part of valor and you should be sure to start by focusing on those things that will end up doing the most good before moving on from there. Don't forget, change for the sake of change won't do anyone any good and will likely serve to create more waste than it will eliminate. Ultimately, it is important to remember that creating a Lean business is a marathon, not a sprint, which means slow and steady wins the race.

Finally, if you found this book useful in any way, a review on Amazon is always appreciated!